THE ORIGIN OF LIFE AND THE INFORMATION PROBLEM

ERIC H. ANDERSON

SEATTLE   DISCOVERY INSTITUTE PRESS   2024

Description

The realization in the twentieth century that even the simplest cells are packed with software tells us something profound about the origin of life. Design theorist and computer programmer Eric Anderson relates the exciting history of the discovery of DNA and shows how the dance of this digital information in each of our cells points insistently away from blind evolution.

Library Cataloging Data

The Origin of Life and the Information Problem by Eric H. Anderson

44 pages, 6 x 9 in.

ISBN-13 Paperback: 978-1-63712-052-1, Kindle: 978-1-63712-054-5, EPub: 978-1-63712-053-8

BISAC: SCI029000 SCIENCE / Life Sciences / Genetics & Genomics

BISAC: SCI027000 SCIENCE / Life Sciences / Evolution

BISAC: SCI075000 SCIENCE / Philosophy & Social Aspects

Publisher Information

Discovery Institute Press, 208 Columbia Street, Seattle, WA 98104

Internet: https://discovery.press/

Published in the United States of America on acid-free paper.

First Edition, First Printing, August 2024.

CONTENTS

The Origin of Life and the Information Problem

Eric H. Anderson

WE SAW IN *The Big Bang and the Fine-tuned Universe* how a series of scientific discoveries in the twentieth century broke strongly against the idea of an eternal universe. Our cosmos had a beginning. And at some point life here on earth had a beginning. How did it happen? Is there any reason to think it was planned and purposeful, or was it just sheer coincidence—the lucky result of some cosmic lottery?

In the final episode of *Star Trek: The Next Generation*, the immortal and nearly omnipotent Q gives Captain Picard the unique opportunity of witnessing the origin of life on Earth. Picard suddenly finds himself standing in a chaotic landscape filled with lava flows and volcanoes that dot the scene. Earth is dark and ominous and utterly devoid of life. As Picard gathers himself from the sudden leap through time and takes stock of his surroundings, Q points excitedly to an oily puddle of chemicals near a volcanic vent.[1]

"Come here," Q says. "There's something I want to show you. You see this? This is you."

Picard gives Q a skeptical glance.

"I'm serious! Right here," insists Q, gazing intently at the chemical sludge. "Life is about to form on this planet for the very first time. A group of amino acids are about to combine to form the first protein—the building blocks of what you call life."

Then, with a playful sneer about the insignificance of humankind, Q says, "Strange, isn't it? Everything you know, your entire civilization, it all begins right here in this little pond of goo."

This memorable exchange between the time-traveling Q and the captain of the Enterprise is of course a work of fiction. But it more or less accurately reflects an idea found in current college textbooks and scientific articles. The idea is this: *If conditions are just right, non-living molecules can give rise to the building blocks of life and, eventually, to life itself. And at some point on the early Earth, the conditions were just right, and voila! It happened. Then, from that first simple life form evolved all the life we now find on Earth, including us. We are the descendants of that first humble organism in that long-ago chemical soup.*

But does the idea hold up to scrutiny? What does the most current evidence suggest for this "goo-to-you" scenario?

Here we are focused on the first part of that story. That is, the claim on the table is that life first emerged through purely natural processes, without any intelligent guidance, intervention, or creative act. The claim is that non-living molecules, by themselves, through nothing more than the laws of physics and chemistry and the random distribution of molecules and chemical reactions, came together to form the building blocks of life, and eventually life itself. This is an idea known as *abiogenesis*.

It's a scientific claim in the sense that we can subject it to scrutiny, testing, and analysis. Of course, we can't travel back in time like Q and Captain Picard to witness the origin of life on Earth. So we can never verify via direct observation the claim that through purely natural processes non-living chemicals turned into organic chemicals and eventually into a living organism on the early Earth. But notice that our ability to investigate any claim about distant historical events is limited in this way, whether it be origin-of-life (OOL) studies, paleontology, archeology, or forensics. Instead, we examine the claim in light of the knowledge and experience we do have, and infer past causes from present clues. In the case of the origin of life, investigators can try to recreate early Earth conditions in the lab. We can run through numerous scenarios with different chemical constituents. We can use our knowledge of chemistry and physics to determine what kinds of reactions would actually be required to produce something like Q's "first protein." We can observe the minimal requirements for the simplest self-reproducing organism alive today and make educated assessments about whether simpler forms of life are possible. We can analyze the many challenges facing a purely natural origin of life and draw reasonable conclusions about whether it is likely to be true.

Assumption Alert

BEFORE WE examine the science, there is one thing we need to be aware of. In addition to the scientific idea of abiogenesis and whatever evidence is marshalled in favor of it, there is an idea or assumption that often hides behind the abiogenesis claim. Here it is in a nutshell: *Even if our current understanding of the origin of life is inaccurate and incomplete, some kind of purely natural process must have generated the first life. Even before we look at the evidence, this must be true. The only question is precisely what that natural process was.*

It is crucial to be aware of this working assumption lurking in the background, to be able to spot it when it hides in the shadows, and to recognize it for what it is. If we want to reason through the question of whether life arose blindly, without intelligent guidance, if we want to see what the physical evidence may be trying to tell us, then we need to set aside this assumption. To cling to the assumption while also looking to investigate the question of whether life first arose through purely natural, unguided processes is akin to trying to investigate whether a house fire was arson or not, but then refusing to consider the possibility that it was arson.

The point may seem so obvious as to be almost unnecessary to make, but many origin-of-life scientists do insist on considering only unguided natural causes for the origin of the first life and, when challenged with contrary evidence, insist that they won't consider anything other

Figure 1. Artist's conception of a primordial landscape with cometary bombardment and a pre-biotic chemical soup in the foreground.

than purely natural causes because supposedly that wouldn't be science. While they insist that they are simply following the evidence, no holds barred, in fact they are barring a possible explanation before they even consider the evidence.

Better to ask ourselves, what does the evidence suggest? Not selective facts chosen to prop up a philosophical position, but the broad array of evidence across the board, the latest and best science we have on the subject, analyzed as carefully and as objectively as we can. What does that kind of careful, objective science say about the origin of life?

Spontaneous Life?

EARLY PHILOSOPHERS and observers of nature, from the Babylonians, to the great Chinese and Indian civilizations, to the ancient Greeks, contemplated the origin of living organisms. How is it that maggots seemed to spontaneously appear in a corpse, or worms from the muddy bank of a river, or even mice from a barrel of wheat? Lacking powerful microscopes and other sophisticated detection equipment, and also lacking in many cases the strong tradition of experiment-based science we take for granted today, early observers could only guess. It seemed as if those creatures arose spontaneously.

It wasn't completely crazy. Some simple observations even seemed to support the idea: get the right conditions, such as the corpse or the mud or the grain, add in the right weather and temperature and, sure enough, over time you were bound to observe maggots and worms and mice. It was an easy mental leap from this observation to the conclusion that such creatures arose spontaneously under the right conditions. As a result, for centuries this idea of spontaneous generation was accepted as the answer to the origin of many forms of life.

Even after the invention of the microscope the idea lived on, but its days were numbered.

Though not the only critic, the great French microbiologist Louis Pasteur (1822–1895) is often credited with the careful experimental approach that finally delivered the death blow to the idea of spontaneous generation. At a time when many scientists still accepted the idea of spontaneous generation, Pasteur performed several experiments with sterilized containers and liquids demonstrating that, when these experiments were carefully performed, living organisms did not arise. Pasteur was later quoted as saying, "Never will the doctrine of spontaneous generation recover from the mortal blow of this simple experiment. There is no known circumstance in which it can be confirmed that microscopic beings came into the world… without parents similar to themselves."[2]

Maxime Schwartz offered the following reflection in the *Journal of Applied Microbiology* on Pasteur's experiments:

> By extremely painstaking experimental methodology, he demonstrated that the appearance of micro-organisms in a presterilized medium could always be explained by germs coming from the outside. He thus succeeded in discrediting any experimental basis for the theory of spontaneous generation.

On a philosophical level, the repercussions were resounding. The onset of life was decidedly not a predictable phenomenon, regularly occurring in any fermentable medium. The question of the origin of life was thereafter clearly set forth—and remains so today.[3]

In other words, if, as Pasteur showed, living organisms normally only come from other living organisms, where then did the *first* organism come from?

A Rose by Another Name

LOOKING BACK, we might be tempted to think that our forebears were simple-minded and foolish. How could they believe in spontaneous generation for so many centuries? After all, every small child today knows that mice don't just arise from wheat, or worms from mud, or maggots from rotting meat. Instead, these creatures come from parent organisms like themselves. We are understandably proud of the great scientific progress that has been made, particularly over the past few hundred years, and we can scarcely comprehend how anyone could not understand what we now take for granted.

Yet if we step away from specific examples like maggots and worms and mice and look at the underlying principle of spontaneous generation more broadly, we are forced to admit that a little more humility is in order. Just as our forebears were wrong about spontaneous generation, is it possible that today's claims of abiogenesis are also in need of careful scrutiny?

Of course, abiogenesis proponents today do not believe in spontaneous generation as it was understood long ago. Yet although the idea of life arising from non-life has been thoroughly discredited by Pasteur and other researchers with respect to the living organisms we observe around us today, what about the beginning of life, what about the first organism? Even if life can't spontaneously arise easily and often, perhaps it could happen at least once, under even more special conditions? The modern abiogenesis story pushes the formation of life from non-life back to a remarkably lucky one-time event in the remote past, yet the core principle remains: *if conditions are just right, non-living matter can turn into living organisms.*

Enter Darwin's Warm Little Pond

IN 1859, just five years before Louis Pasteur rejected the idea of spontaneous generation at a scientific conference, another European, Charles Darwin, published his formidable work that would become a landmark in biology, *On the Origin of Species.*

Darwin's book did not attempt to address the origin of life. He simply assumed one or more original self-reproducing organisms, and built his theory of evolution from there. Beginning in later editions he does mention "the Creator" as the possible source of the first living organism or organisms, but he apparently held out hope for a purely naturalistic source for the origin of the first life. For even after Pasteur proved that spontaneous generation wasn't happening all around us, Darwin privately offered an explanation for the origin of life that didn't involve a creator.

"How on earth is the absence of all living things in Pasteur's experiment to be accounted for?" Darwin asked his friend Joseph Hooker in an 1871 letter. Darwin went on from there to speculate: "It is often said that all the conditions for the first production of a living organism are now present, which could ever have been present.— But if (& oh what a big if) we could conceive in some warm little pond with all sorts of ammonia & phosphoric salts,—light, heat, electricity &c present, that a protein compound was chemically formed, ready to undergo still more complex changes...."[4]

Darwin immediately acknowledged that in the present natural environment such a "protein compound" would be "instantly devoured, or absorbed." But perhaps it wasn't always this way? Perhaps before living things were around, he mused, before life had formed on the early Earth, a protein compound could have survived and continued to evolve—undergoing, as Darwin suggested, "still more complex changes."

Notice that Darwin wasn't proposing that life can easily arise from non-living matter, or that the process often occurs. Rather, he was asking, what if, perhaps under just the right conditions, an early precursor to life could arise from non-living chemicals? And if so, might not that precursor eventually lead to a living organism?

Darwin was not alone in his musings about the possibility of chemicals coming together in a blind shuffle to form a living organism. In the decades that followed, a tremendous amount of effort would be spent trying to flesh out the idea and provide some experimental support.

Chemical Soup, Anyone?

IN THE early 1920s, the great Russian biochemist Alexander Oparin proposed that life originated on Earth through a series of biochemical steps. Oparin thought that simple organic molecules could form on the early Earth in an atmosphere containing methane, ammonia, hydrogen, and water vapor, but one in which, unlike our present atmosphere, oxygen was absent or significantly minimized, preventing oxidation of other chemicals in the environment. (Such an oxygen-starved atmosphere is known as a reducing atmosphere.) Oparin speculated that after these simple organic molecules had formed under such conditions, occasionally they would react to form more complex molecules, developing new properties and eventually turning into a living organism.[5]

Shortly thereafter, British geneticist J. B. S. Haldane proposed that the primitive ocean on the early Earth was like a "hot dilute soup" in which simple organic compounds could have formed.[6] Like Oparin, and Darwin before, Haldane thought the simple compounds would react with others, forming more complex compounds, then the components of living cells, and eventually a living organism.

Although their ideas were initially dismissed by many scientists, Oparin's and Haldane's suggestions that life started on Earth through a kind of "chemical evolution" gained traction. Some scientists remained skeptical, but to many it seemed that it was only a matter of time before the case was clinched for this scenario. After all, if Darwin had shown that all the wonderful complexity and diversity of life forms on our planet could evolve from that first simple organism, through a process of natural selection acting on random variations without the need for any guiding intelligence, then perhaps that first living organism also could be explained as the result of a purely natural process. Perhaps nature itself was the creator—producing from the simplest chemical elements the building blocks of life and, eventually, life itself.

This possibility continued to tantalize the biochemists of the day. Yet while the theory sounded good, what was still needed was hard experimental evidence. Enter Stanley Miller and Harold Urey.

A Personal Journey

WHEN I first learned about the famous experiments performed by Stanley Miller and Harold Urey at the University of Chicago,[7] I was impressed as I heard about the formation of amino acids from nothing but simple compounds and a bit of electricity—amino acids that, I was told by abiogenesis proponents, could have come together to form simple proteins and eventually more complex organic molecules, and eventually life, just as Q told Captain Picard and just as Darwin speculated.

Was it true? Rather than just accepting what I was being told, I decided to find out for myself. As I read more about the origin of life and the often heated debate over the possibility of undirected chemical reactions producing a primitive living organism, I wondered why there was still so much debate if Miller and Urey had shown that life, or at least the building blocks of life, could arise from a purely natural process.

After all, the preeminent scientist George Gaylord Simpson had noted fully sixty years ago that "at a recent meeting in Chicago, a highly distinguished international panel of experts was polled. All considered the experimental production of life in the laboratory imminent."[8] So, if nearly 70 years ago, Miller and Urey had made such a big breakthrough, and if at the turn of the next decade, in 1960, a distinguished panel of

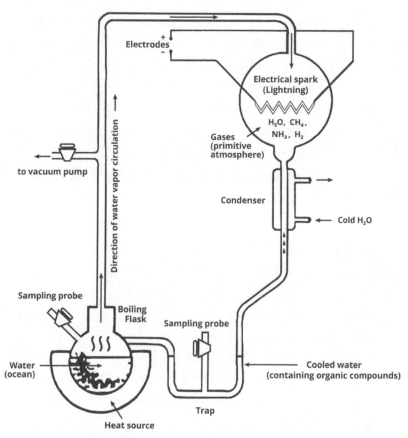

Figure 2. Rendering of the setup used in the famous Miller-Urey experiments.

experts confidently affirmed that laboratory confirmation for the purely natural origin of life was just around the corner, what was there still to debate all these decades later?

I wondered if perhaps the debate raged on because some people believed that a naturalistic origins story would conflict with their personal religious or philosophical views about a purposeful creation. Maybe the researchers had sealed the deal decades ago, and the religious folks just weren't willing to face the music. Yet as I researched the topic, I found that many of the scientists and authors who were critical of the Miller-Urey experiment raised questions based not on any religious viewpoint, but on the science. Indeed, one of them, Dean Kenyon, had been a leading proponent of a naturalistic origin of life, and had literally written the book on the topic—or to be precise, had co-authored a leading origin-of-life textbook, *Biochemical Predestination*.[9]

I myself didn't have a philosophical or religious concern with the abiogenesis story, but something about it smelled fishy. And the more I researched it, the fishier it smelled.

To be sure, Miller and Urey had done wonderful science and created a very ingenious experiment. They were at the forefront of research in the early 1950s. Through their hard work and dedicated efforts, we learned new and important things about how non-living chemicals behave and how certain organic compounds might be formed. But as I investigated further, what I found was this: the idea that the now-iconic Miller-Urey experiment placed us firmly on the path to explaining a purely natural origin of life—an idea pushed primarily, I might add, by people other than Miller and Urey—is simply false.

In the decades since, nearly every aspect of the Miller-Urey experiment has been challenged. It taught us interesting things, but it fell far short of replicating a mindless chemical evolutionary process or the relevant conditions likely to have held sway on the early Earth. From the reducing atmosphere used in the experiment,[10] to the need for just the right amount of energy,[11] to the careful isolation of the tender chemicals from unfavorable interfering cross-reactions,[12] to the protective environment in which the reactions took place, careful observers have questioned the relevance of the Miller-Urey results to the origin of life on the early Earth.[13] Yet the Miller-Urey experiment is still touted in many high school and college textbooks as proof that the formation of the chemicals necessary for life on the early Earth is no longer a serious problem and has been largely solved.[14] Nothing could be further from the truth.

And that isn't even half the problem for abiogenesis.

Flash in a Flask

EVEN IF we were to accept the inaccurate textbook story of the Miller-Urey experiment as gospel, we would not be justified in concluding that it showed a viable pathway to a naturalistic origin of life. That's because, at the end of their carefully controlled, intelligently designed and intelligently guided experiment, they were still light years away from getting a simple living organism.

There are so many additional problems with the overall abiogenesis story that it is hard to know where to begin. Researchers have identified more than a dozen serious problems with the abiogenesis account. Most of them, even individually by themselves, doom the abiogenesis idea. When taken together, they constitute a devastating critique of the naturalistic origin-of-life story. In 1982, organic chemist and molecular biologist A. G. Cairns-Smith raised several objections against the typical "origin of life" simulation experiments.[15] Soon thereafter, chemist Charles Thaxton, geochemist Roger Olsen, and materials scientist Walter Bradley provided a rigorous critique of the many origin-of-life proposals and speculations. They referred to a "crisis in the chemistry of origins" and observed that "the undirected flow of energy through a primordial atmosphere and ocean is at present a woefully inadequate explanation for the incredible complexity associated with even simple living systems, and is probably wrong." They go on to conclude that "reasonable doubt exists whether simple chemicals on a primitive earth did spontaneously evolve (or organize themselves) into the first life."[16]

The situation has not improved since that time. Quite the contrary. Each additional avenue of research seems to spawn additional questions and challenges to the abiogenesis story. Researchers' abiogenesis ideas are regularly shot down by other origin-of-life researchers, who then go on to propose their own, equally inadequate suggestions.

A recent 2019 paper examines multiple proposed locations for the origin of life, including Darwin's "warm little pond," hot springs, outer space, and (a popular suggestion nowadays) deep-sea hydrothermal vents. The researchers conclude that none of these locations are able to meet the requirements for abiogenesis, and instead propose a geyser system "driven by a natural nuclear reactor."[17] In a recent review article on origin-of-life research, astrobiologist and theoretical physicist Sara Walker lauds the efforts that have been made in origin-of-life research to date, but acknowledges that "we have not yet been able to answer the question of how life first emerged."

Walker appears unwavering in her faith that the first life on earth arose naturalistically, but after examining problems with many current attempts to address the origin of life, she concludes that "novel approaches... may be required" and hopes for a "new theory of physics" that can help bridge the gap. Walker comments that the task of understanding how life arose from purely natural causes might be as difficult as "unifying general relativity and quantum theory," and suggests that solving the puzzle of our origins might occur only "if we are so lucky as to stumble on [a] new fundamental understanding of life."[18]

Thus have the days of early excitement over the Miller-Urey experiment been replaced by an understanding of a most sobering reality. In the remainder of this mini-book and in the next, we will review just two of the key problems with the modern abiogenesis story: the need for biological information and the challenge of self-replication.

More Information, Please

THE EARLY 1980s were a time of great excitement in the computer world. Just a few years earlier Apple Computer Company had been founded by Steve Jobs, Steve Wozniak, and Ronald Wayne, kicking off a revolution of affordable personal computers that began making their way into the hands of hobbyists, computer clubs, and a few early homes. Apple's popularity had exploded with the introduction of the Apple II in 1977. Several other manufacturers also had started developing and selling computers directly to consumers, with names like Altair, Texas Instruments, TRS-80, Sinclair, Atari, Commodore, and others.

My father was an engineer by training and had more than a passing interest in the young field of personal computing. As soon as circumstances and the family budget allowed, he loaded my three brothers and me into the car early one Saturday morning for an eight-hour drive to a computer fair where we pored over the thrilling new offerings in this young field. Ignoring our exhaustion from a busy day, we talked excitedly the whole drive home about the new technology we had seen.

Although my father was a frugal man, he had done his research and was determined to spend his hard-earned funds on a quality machine that could be used for years to come. After careful analysis of the pros and cons and the costs and benefits of various systems, at a level of detail that only an engineer could appreciate, he settled on one of the better Apple II-compatible systems he could find at the time. With my mother's eventual agreement that this would take the place of all our gifts that Christmas, he reached deep into his pocketbook and placed the order.

Looking back, I have to smile at our first computer, one that cost more than many high-end gaming systems today. When we finally got our new computer a few weeks later on Christmas Eve, we could scarcely wait until Christmas morning to set it up. It had all the bells and whistles! A whopping 64K of RAM (not the 48K that most of the Apple IIs came with at the time), a 5.25" floppy disk drive, five video games, ten blank floppy disks, a simple joystick, and, best of all, a large CRT *color* monitor! (No way we were going to settle for a monochrome green monitor!)

Within a few months, we added a dot matrix printer and a second drive. Now we had two—count 'em, two!—5.25" floppy disk drives, which enabled the computer to run more advanced word processing programs and also allowed us to copy disks much more easily.

Although our family did not even own a television, we were soon the technological envy of our friends and acquaintances. We had the first personal computer of anyone in the neighborhood. Suddenly our garage, which had earlier been converted into a family play area, became the scene of countless weekends and late nights as my siblings and I and our friends huddled around that low-resolution color monitor playing exciting round after round of early 8-bit computer games. To my parents' chagrin, my sleeping habits took a serious turn for the worse as the afternoons at the computer turned into evenings, then nights, then early mornings.

But it wasn't all games. My father's engineering background instilled in us a desire to not just use the technology, but to understand how things worked.

I purchased an early manual on the Apple operating system and studied the details for hours. I learned how to "hack" into computer games and change some of the game parameters and screen displays. Not a particularly valuable thing in its own right, but in the process I learned valuable lessons about file systems, disk sectors, storage protocols, and other inner workings of the computer.

I also taught myself to program in BASIC, the simple integer language used by the Apple II, and began to write my own programs, eventually doing some early database work. Soon I delved into COBOL and Fortran, and even spent a couple of summers helping program an old Burroughs computer in Hexadecimal—now that was tedious work! Eventually I took to building my own computers.

Looking back on this formative time, both in the computer industry and in my own life, I am grateful for the opportunity I had, in my own small way in my own small corner of the world, to witness firsthand and up close and personal the remarkable transformation that computers and information technology would come to play in the world. Perhaps the biggest takeaway from the experience was the realization that information was key. It wasn't the metal or the plastic or the wires or the magnetic disks. Yes, those were important. But it was the information, both in the subtle way the physical parts were arranged for a particular purpose, and more obviously in the way the codes and the programs brought those careful arrangements of parts to life. It was always the same thing at the heart of every game, every database, every floppy disk sector, every function: information.

In the world of complex functional systems, information is king. What does that have to do with the origin of life? As it turns out, an important form of digital information was around long before computers. Information sits—or rather hums and dances—at the heart of all life.

A Strand of DNA Walks into a Bar...

THE STORY is well known in scientific circles, almost the stuff of legend. One Saturday in February 1953, English scientist Francis Crick walked into *The Eagle* pub in Cambridge, England, with his American research partner James Watson, and announced that the two "had discovered the secret of life!"[19]

Watson and Crick might be forgiven for their bravado. After all, they had, along with important assistance from Rosalind Franklin and Maurice Wilkins, discovered the structure of the large organic molecule arguably most central to life: deoxyribonucleic acid (DNA).

Determining the three-dimensional helical structure of this important molecule was achievement enough, but it also occurred to Watson and Crick that the pairing of nucleotides across the twisting ladder of DNA suggested "a possible copying mechanism for the genetic material."[20] That is, the structure of DNA might facilitate the copying of genetic information from generation to generation. They turned out to be spectacularly correct.

A remarkable glimpse into history has been preserved in the form of a handwritten letter Crick penned to his 12-year old son, Michael, on March 19, 1953.[21] Despite writing to a young man not yet in his teens, Crick lays out the structure of DNA in some detail:

My Dear Michael,

Jim Watson and I have probably made a most important discovery. We have built a model for the structure of de-oxy-ribose-nucleic-acid (read it carefully) called D.N.A. for short....

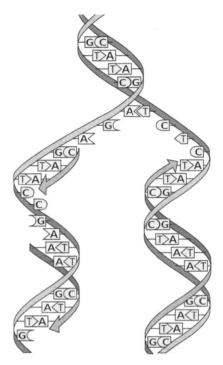

Figure 3. DNA structure, showing the nucleotide base pairs
adenine and thymine (A/T) and cytosine and guanine (C/G), and
the double helix unwinding as part of the copying process.

Now on one chain, as far as we can see, one can have the bases in
any order, but if their order is fixed, then the order on the other chain
is also fixed. For example, suppose the first chain goes A-T-C-A-G-T-T,
then the second must go T-A-G-T-C-A-A.

Crick went on to suggest that this complementary pairing of the two
DNA strands not only shed light on the information-bearing properties
of DNA and the existence of what would later become known as the
"genetic code," but also hinted at a potential copying mechanism:

It is like a code. If you are given one set of letters you can write down
the others.

Now we believe that the D.N.A. is a code. That is, the order of the
bases (the letters) makes one gene different from another gene (just as
one page of print is different from another)....

In other words we think we have found the basic copying mechanism by which life comes from life…. You can understand that we are very excited. We have to have a letter off to *Nature* in a day or so. Read this carefully so that you understand it. When you come home we will show you the model.

As I ponder Crick's letter and imagine what it must have been like for Crick and Watson to discover the structure of DNA, I am impressed by their insight into the implications as they recognized not only the structure of DNA but also the existence of a code, the key role of information, and the way "life comes from life," as Crick explained to his son.

Just a few years later, in 1957, Crick gave a lecture in which he outlined what he called the "sequence hypothesis."[22] The following year he published a paper titled "On Protein Synthesis."[23] Among other details, he proposed that the "specificity" of a piece of DNA (meaning, in essence, the information contained in that piece of DNA) results from the sequence or the ordering of the base molecules or bases. In other words, it isn't merely the physical structure or the chemical makeup of the DNA bases that store information, but the arrangement of those bases.

Like the letters of the English alphabet you are reading in this book, it is the *order* of the letters that conveys the information, not the color of the ink or the type of paper used.

At least in this regard, Crick's sequence hypothesis was spot on. Watson and Crick's work stands as a singular achievement in the history of science. "Watson and Crick's discovery would forever change our understanding of the nature of life," observed philosopher of science Stephen Meyer. "At the close of the nineteenth century, most biologists thought life consisted solely of matter and energy. But after Watson and Crick, biologists came to recognize the importance of a third fundamental entity in living things: information."[24]

The Language of Life

SINCE THAT fateful day in the English pub in 1953, the problem of explaining the origin of life has increasingly been recognized as a problem of explaining the origin of biological information. Where did the information in life come from? More than sixty years later, scientists are still asking the question, even as they seek to better understand molecular life.

Organic life, it is now clear, is all about nano-technology, molecular machines, information-processing systems and, of course, DNA. As we peer into ever more sophisticated microscopes, it becomes increasingly clear that we are dealing with a carefully orchestrated process complete with a 4-bit digital code, storage, retrieval and translation mechanisms, computation protocols, and other hallmarks of a highly functional, information-rich system.

How did such a system first arise? As we noted, we were not around to see the first living organism appear on the early Earth and we don't have the luxury of *Star Trek's* Q to transport us back in time to witness it. But are there clues that can help us determine the origin of these living systems?

There are at least two things we know about functionally integrated, information-rich systems like those we find throughout biology—two things that can help us draw a reasonable conclusion about the origin of such systems.

First, there is no known naturalistic cause that has been shown to produce large amounts of novel information. Such a thing has never been observed. Not even once. And there is no theory successfully detailing how it might happen. Moreover, it is not as though scientists have proposed several great possibilities that just haven't quite panned out, or that several good naturalistic explanations are on the table and just need a little more tweaking or a little more research or a little more funding. Although natural processes can be called upon to explain many observations, when we are dealing with functionally integrated information-rich systems, the situation is completely different. It is not as though the search has only recently begun, we are very close, and we just need to try a little harder or look a little longer. The search has been carried out extensively over multiple generations by thousands of researchers through hundreds of thousands of hours of effort and at a cost of billions of dollars, without success. No credible naturalistic process has ever been identified that can produce information-rich systems.

Renowned synthetic organic chemist James Tour has analyzed what he sees as the most promising recent origin-of-life (OOL) research efforts. An experienced researcher with over 700 research publications and more than 130 patents to his name, he has a knack for seeing through the optimistic hype and focusing on the real-world difficulties of assembling functional molecular systems. In a 2019 critical essay, Tour expressed frustration over the lack of candor in many origin-of-life papers about the state of the field, and he asked, "Why not admit what we cannot yet explain: the mass transfer of starting materials to the molecules needed for life; the origin of life's code; the combinatorial complexities present in any living system; and the precise nonregular assembly of cellular components?" Wishful thinking, or perhaps naivete, may also play a role. "I have discussed these issues with OOL researchers," he writes, "and I am amazed that they fail to appreciate the magnitude of the problem in building molecules."[25]

Actually, the situation is even more desperate than that.

It is increasingly apparent that chance and law-like processes not only cannot *build* information-rich systems, but actually *destroy* information. On a net basis, these natural processes invariably result in a *loss* of information over time. This is a well-understood principle, observed over and over in the field. It is clear from mathematical analyses and principles of information theory. It is confirmed from lab research.

The conclusion to draw from these various observations is straightforward: processes that are either driven by a law-like cause or that come about haphazardly by chance events simply do not have the capacity to generate new information-rich systems such as those we see in biology. Based on our current understanding, we have good reasons for concluding that a purely natural explanation for these kinds of systems will *never* be found.

Indeed, in an ironic twist, it turns out that organisms have systems in place, including proofreading and error-correction mechanisms, geared specifically to combat the information-destroying tendencies of law and chance. The origin of these sophisticated proofreading and error-corrections systems also requires explanation. For someone to then turn around and suggest that law and chance produced these information-rich systems in the first place is completely upside down and backwards.

This brings us to the second thing we know. We know of only one type of cause with the demonstrated power to produce functionally integrated, information-rich systems of this sort. That cause is intelligence.

Examples abound. All around us we have desktop computers, smartphones, cloud servers, smart TVs, web-based applications, connected devices, and on and on. All of these functionally integrated systems—billions and billions of them around the world—and the information contained within them came about through a process of goal-oriented activity, preparation, planning, gathering requirements, choosing materials, creating prototypes, identifying interconnectivity protocols, and so forth. In other words, they came about through a purposeful design process, through the activity of mind. Not a single one of them came about by purely natural causes.

Even simpler information-rich artifacts, such as a book or an Instagram post—all of these trace back to a mind, an intelligence.

As Meyer has noted, "Our uniform experience affirms that specified information—whether inscribed in hieroglyphics, written in a book, encoded in a radio signal, or produced in a simulation experiment—*always* arises from an intelligent source, from a mind and not a strictly material process.... Indeed, whenever we find specified information and we know the causal story of how that information arose, we always find that it arose from an intelligent source."[26]

And note that this is a conclusion based not on ignorance or idle speculation or lack of knowledge, but on the many things we have learned and observed in molecular biology and elsewhere—on what we *know*. It is a conclusion based on our regular, repeated, and uniform experience and observations.

Whodunit?

No SCIENTIST was around at the time life was first formed on the Earth. There were no bloggers or reporters or cameras or video blogs to record the event or give us a "breaking news" report on how life started. Q from *Star Trek* can't take us back for a front-row seat of the main event. The origin of life was a historical event, one that took place long ago, far in the past and beyond our ability to witness. As a result, and as with all of the historical sciences, the way forward is to assemble clues, consider competing explanations, and carefully determine which proposed cause is the most likely, the one that most adequately explains the clues on the table.

Consider an illustration. Suppose you've recently moved to Eastern Washington from out of state, and while putting in a garden in your backyard you come across a layer of white powdery material a little way below the surface. You later find out that the same layer can be found throughout the region. You quickly run through some possible causes. A flood? A disastrous outpouring of pollutants from an old factory in town, one long ago closed down? A volcano? Each of these causes can impact a large geographic area and will affect the geology of a region. However, you quickly recognize that only one of the proposed causes has the demonstrated capacity to actually produce the effect in question. A big flood might lay down a layer of silt over a wide area, but not ash. An industrial disaster could not lay down the sheer volume of material you've encountered. Plus, industrial disasters don't produce volcanic ash, and after getting a chemist friend to run some tests, you confirm that it indeed is volcanic ash. So you draw a reasonable inference from the evidence: the white powdery ash was produced by a volcanic explosion.[27]

Notice how with careful reasoning and analysis, you can draw a reasonable inference from the physical evidence and come to the right conclusion, even if you had never heard about or witnessed that fateful Sunday morning in the spring of 1980 when Mount St. Helens erupted, sending a massive plume of ash heavenward that eventually spread as far as Canada.

As we consider the origin of life, we are like a scientist examining a deposit of ash without having been there to actually see the volcanic eruption. Because we weren't there to observe the origin of life, we can't replay a video of the event and say, "There it is. That's how it happened." But that doesn't mean we are reduced to unsupported speculations, or to throwing up our hands and giving up on the question. Instead, we can look at the clues and the evidence and draw upon our own experience to come to a reasonable conclusion—what philosophers of science call an inference to the best explanation.

In the case of the origin of life, what needs explaining isn't a layer of volcanic ash, but instead the presence of information in the first living cell. And the cause is clear, if we are but willing to entertain it. All of the many naturalistic explanations that have been proposed over the years fail to explain the information-rich systems we see in even the simplest single-celled organism. Indeed, naturalistic processes tend to degrade information over time, not create novel information. As a result, we can reasonably conclude that none of these explanations can be what philosophers of science call the "true cause," because none of them has the capacity to produce what we observe in molecular biology. However, we do know of one cause that regularly produces large amounts of information and repeatedly produces complex, functionally integrated, information-rich systems, such as we find in living cells. That cause is intelligence. Unlike the naturalistic explanations, creative intelligence can be the true, the adequate cause, because it is capable of producing these kinds of effects.

So the clear and reasonable conclusion we can draw for the origin of life and the information-rich systems we find in even the simplest organism—the inference to the best explanation—is that they were caused by the activity of an intelligent agent.

A Vanishing Act

SOME PROPONENTS of the naturalistic abiogenesis story counter this information argument to design by attempting to explain away the information in DNA. In essence they try to make the information vanish. They employ either of two strategies to this end.

First, some have tried to explain away the information-rich content of DNA by arguing that there actually is no information in DNA. We might think there is information in DNA, but it isn't really there, they argue. We, as humans, are simply imposing our own biases and expectations on DNA, and really it is just a molecule like any other. What we call "information" is simply our way of understanding the molecular makeup and behavior of DNA.

If this seems to you like a strange argument, you are not alone. When Watson and Crick discovered the structure of DNA and recognized the complementary base pairing and copying mechanism, they were not just imposing their own expectations and biases on the molecules. They were discovering something that actually existed in the real world, something that was already there long before they were born and before they turned their attention to the cellular world. They certainly had no religious bias prodding them to see information in the cell. Neither man was particularly religious, both held out hope that a purely materialistic explanation for the origin of life might one day be found, and Crick even went so far as to suggest that the first life on Earth was seeded from outer space, having originated on some faraway planet. This idea only backs up the problem to another planet, and it adds the additional difficulty of safely transporting microscopic life across untold millions of miles of cold space while it's being bombarded by harmful cosmic rays. But the fact that he was willing to reach for such an explanation clearly shows that this was not a man eager to acknowledge purposeful design as an explanation for the origin of life on Earth. And yet he and Watson readily acknowledged that the cell is rich with information and information-processing systems.

Subsequent discoveries have only confirmed and deepened that understanding. Thus, it has been known for decades that there is information in DNA—real, observable, encoded, functional information. Numerous successful companies have been founded to retrieve, study, and analyze the information in DNA. University biology departments and even some computer departments have created courses to teach the next generation of scientists about the information in DNA. An entire field of research called "bioinformatics" has arisen in recent years, dedicated to the study of information in biological systems, in particular the information in DNA.

Other opponents of design in biology have taken just the opposite tack, arguing that there is information in everything. The assumption behind this argument is that all matter in the universe contains information—the rocks, the particles, the stars, and the galaxies. Sure, DNA contains information, they argue, but so does everything else. So there is nothing unique about DNA. Nothing to see here, folks. Move along.

One person I've debated argued, for example, that DNA is just a molecule like any other molecule and that there is nothing more interesting in DNA than in a glass of salt water, because, as he claimed, there is information in both DNA and salt water.

This, too, is nonsense. Even a child can readily understand that there is a world of difference between what we find in DNA and what we see in something like a glass of salt water or the particles of lifeless matter floating about the universe.[28] It's the same difference as between the letters in a meaningful book and page after page of random characters (in the case of particles floating haphazardly about the universe), or (in the case of salt water) page after page of a relatively simple pattern of repeating letters—for example, aabbccdd, aabbccdd, aabbccdd repeating over and over.

In contrast to these non-living examples, what we find in DNA is neither a simple repeating pattern nor a random collection of nucleotides. Instead, DNA contains highly specified, functional information stored directly in digital form and expressed through the genetic code.

How did the information we find in cells arise in the first place? The question persists. Any attempt to ignore it or to sweep it under the rug by denying or trivializing the existence of biological information is intellectually empty and gives us a hint that perhaps there is some other agenda than open-minded scientific inquiry at play.

Coming Full Circle

TODAY, WITH the tremendous developments over the last few decades in our understanding of DNA and the genetic code, along with the discovery of ingenious molecular machines at the cellular level, we are better positioned than ever before to answer the question at the beginning of this mini-book: How did life begin? The accumulated evidence we have today points toward a planned, purposeful, carefully designed origin of life as the best explanation.

There are many origin-of-life researchers who reject this conclusion, but they do so primarily by insisting dogmatically that scientists must only entertain purely naturalistic, mindless processes to explain the origin of life. Their claim that life arose on the early Earth through a long series of unguided chemical interactions is hardly more valid than the distant speculations of our ancient forebears about spontaneous generation. Actually, it's arguably less valid because modern materialists no longer have the excuse of ignorance. People once thought that life routinely emerged from non-life. We now know better. And no further back than Darwin's time many scientists assumed that microscopic life was fairly simple. Now we know that even the simplest cell is a marvel of information processing and engineering sophistication.

Despite Darwin's creative conjectures about the first cell, despite the theories of Oparin and Haldane, and the careful experiments of Miller and Urey, despite the subsequent decades of discovery—and indeed, in large part because of those discoveries—it's clearer than ever that something more than time and blind material processes are required to conjure up the first living cell.

When we step away from the vague proposals and speculations from the past and examine what is actually required for a living organism—what is actually required to assemble a complex, functional, information-rich organism—the answer is clear. The first life was not an accident of chemistry or a lucky draw of the cosmic lottery. Rather, life was intended. It was planned. It was orchestrated.

It was designed.

Review: Your Turn

1. How do scientists investigate and reason about past events that they can never directly observe, such as the origin of the first life on Earth?

2. Why is the assumption that life must have started by purely natural processes, without any intelligent guidance or intervention, a philosophical assumption rather than a scientific one?

3. Instead of assuming that life must have originated by purely natural processes, what evidence can we consider to determine whether this is true or whether life required a creative intelligence?

4. Did the Miller-Urey experiment show that life could arise on the early Earth by natural causes? What did it show?

5. What does DNA contain that makes it so different from non-living matter?

6. Even the simplest cells are brimming with information. What type of cause has the demonstrated ability to generate new information?

FUEL YOUR CURIOSITY!

Recommended Resources for Further Exploration:

VIDEOS

 The Information Codes Inside Your Body

 Information Enigma: Where does information come from?

 The MYTH of Junk DNA

PODCASTS

 Physicist Eric Hedin: Information, Entropy, First Life

Meyer & Tour on New Critiques of Origin of Life Research

On the Origins of Life
David Berlinski

On Origin of Life, Chemist James Tour Has Called These Researchers' Bluff
Brian Miller

evolutionnews.org
Original reporting and analysis about evolution, neuroscience, bioethics, intelligent design and other science-related issues, including breaking news about scientific research.

discovery.org/id/
The institutional hub for scientists, educators, and inquiring minds who think that nature supplies compelling evidence of intelligent design.

intelligentdesign.org
Documents the mounting scientific evidence for nature's intelligent design. Through this site, you can explore the evidence for intelligent design for yourself.

ENDNOTES

1. *Star Trek: The Next Generation*, season 7, episode 26, "All Good Things," aired May 23, 1994. Excerpt at YouTube, video, 0.46, https://www.youtube.com/watch?v=YLyqTtrhUJE.

2. B. Lee Ligon, "Biography: Louis Pasteur: A Controversial Figure in a Debate on Scientific Ethics," *Seminars in Pediatric Infectious Diseases* 13, no. 2 (April 2002): 134–41.

3. Maxime Schwartz, "The Life and Works of Louis Pasteur," *Journal of Applied Microbiology* 91 (October 2001): 598, https://sfamjournals.onlinelibrary.wiley.com/doi/epdf/10.1046/j.1365-2672.2001.01495.x.

4. Darwin to J.D. Hooker, February 1, 1871, DCP LETT 7471, Darwin Correspondence Project, University of Cambridge, https://www.darwinproject.ac.uk/letter/DCP-LETT-7471.xml. The historical context, as well as additional references in Darwin's letter to Hooker, underscore that speculation about life arising from non-life was not just an isolated musing or offhand remark by Darwin, but was something being actively discussed in the larger scientific community at the time.

5. See for example, "Aleksandr Oparin," *Encyclopedia Britannica*, April 17, 2019, accessed February 10, 2020, https://www.britannica.com/biography/Aleksandr-Oparin.

6. J. B. S. Haldane, "The Origin of Life," *The Rationalist Annual* 148 (1929): 3-10; reprinted in J. B. S. Haldane, *Science and Life: Essays of a Rationalist* (London: Pemberton, 1968).

7. Stanley L. Miller, "A Production of Amino Acids Under Possible Primitive Earth Conditions," *Science* 117 (May 1953): 528–9.

8. George Gaylord Simpson, "The World into Which Darwin Led Us," *Science* 131 (1960): 966–74.

9. Dean H. Kenyon and Gary Steinman, *Biochemical Predestination* (New York: McGraw-Hill, 1969).

10. See, for example, Sidney W. Fox and Klaus Dose, *Molecular Evolution and the Origin of Life*, rev. ed. (1972; repr., New York: Marcel Dekker, 1977), 43; Freeman Dyson, *Origins of Life*, 2nd ed. (Cambridge: Cambridge University Press, 1999), 33–34; and David C. Catling, "Comment on 'A Hydrogen-Rich Early Earth Atmosphere,'" *Science* 311 (2006), author reply 38, https://doi.org/10.1126/science.1117827.

11. See J. P. Ferris and D. E. Nicodem, "Ammonia: Did It Have a Role in Chemical Evolution?" in *The Origin of Life and Evolutionary Biochemistry*, eds. K. Dose, S. W. Fox, G. A. Deborin, and T. E. Pavlovskaya (New York: Plenum Press, 1974), 107; see also a discussion of energy factors impacting the hydrothermal vent hypothesis in J. Baz Jackson, "The 'Origin-of-Life Reactor' and Reduction of CO_2 by H_2 in Inorganic

Precipitates," *Journal of Molecular Evolution* 85, no. 1–2 (2017): 1–7, https://doi.org/10.1007/ s00239-017-9805-9.

12. Robert Shapiro, "Prebiotic Cytosine Synthesis: A Critical Analysis and Implications for the Origin of Life, *PNAS* 96, no. 8 (April 1999): 4397–98.

13. See Fox and Dose, *Molecular Evolution and the Origin of Life*, 74–76; and Robert Shapiro, *Origins: A Skeptic's Guide to the Creation of Life on Earth* (New York: Summit Books, 1986), 112.

14. In his book *Icons of Evolution: Science or Myth?* (Washington, DC: Regnery Publishing, 2000), biologist Jonathan Wells reviews in detail the persistence of textbooks overselling the significance of the Miller-Urey experiments. For an updated discussion, see Charles B. Thaxton et al., *The Mystery of Life's Origin: The Continuing Controversy* (Seattle: Discovery Institute Press, 2020), chap. 16.

15. A. G. Cairns-Smith, *Genetic Takeover and the Mineral Origins of Life* (New York: Cambridge University Press, 1982).

16. Charles B. Thaxton, Walter L. Bradley, and Roger L. Olsen, *The Mystery of Life's Origin* (New York: Philosophical Library, 1984).

17. Shigenori Maruyama et al., "Nine Requirements for the Origin of Earth's Life: Not at the Hydrothermal Vent, but in a Nuclear Geyser System," *Geoscience Frontiers* 10, no. 4 (2019): 1337–57, https://doi.org/10.1016/j.gsf.2018.09.011.

18. Sara Imari Walker, "Origins of Life: A Problem for Physics, a Key Issues Review," *Report on Progress in Physics* 80, no. 9 (August 2017), https://doi.org/10.1088/1361-6633/aa7804.

19. Recounted by James D. Watson in *The Double Helix: A Personal Account of the Discovery of the Structure of DNA* (New York: Touchstone, 2001). See also Bill Mesler and H. James Cleaves II, *A Brief History of Creation: Science and the Search for the Origin of Life* (New York: W. W. Norton, 2016), 199–200.

20. James D. Watson and Francis H. C. Crick, "A Structure for Deoxyribose Nucleic Acid," *Nature* 171 (April 1953): 737–38.

21. Jane J. Lee, "Read Francis Crick's $6 Million Letter to Son Describing DNA," *National Geographic Society Newsroom*, National Geographic Society, April 11, 2013, https://blog.nationalgeographic.org/2013/04/11/read-francis-cricks-6-million-letter-to-son-describing-dna/.

22. For a discussion of Crick's remarkable 1957 lecture delivered at a symposium of the Society for Experimental Biology, at University College London, see Matthew Cobb, "60 Years Ago, Francis Crick Changed the Logic of Biology," *PLOS Biology* 15, no. 9 (September 18, 2017): e2003243, https://journals.plos.org/plosbiology/article?id=10.1371/journal.pbio.2003243.

23. Francis H. Crick, "On Protein Synthesis," *Symposia of the Society for Experimental Biology* 12 (1958): 138–63.

24. Stephen C. Meyer, *Signature in the Cell: DNA and the Evidence for Intelligent Design* (New York: HarperOne, 2009), 84.

25. James Tour, "Time Out," *Inference: International Review of Science* 4, no. 4 (July 2019), https://inference-review.com/article/time-out.

26. Meyer, *Signature in the Cell*, 347.

27. Example drawn from Meyer, *Signature in the Cell*, 342. See also "In a Three-Way Radio Debate, Stephen Meyer Takes on a Chemist and a Biologist," March 16, 2016, in *ID the Future*, podcast, MP3 audio, 27:55 (starting at 6:34), www.discovery.org/multimedia/audio/2016/03/in-a-three-way-radio-debate-stephen-meyer-takes-on-a-chemist-and-a-biologist/.

28. For an in-depth analysis of several incorrect arguments often put forward by opponents of intelligent design regarding information in biology, see my series of interviews at *ID the Future*, podcast, MP3 audio, www.discovery.org/multimedia/?s=eric+anderson.

IMAGE CREDITS

Figure 1. Primordial landscape. "Chemical Soups around Cool Stars." Illustration by NASA/JPL-Caltech. Public domain.

Figure 2. Rendering of the setup used in the Miller-Urey experiment. Adapted by Brian Gage from various images, including image by Yassine Mrabet, 2008, Wikimedia Commons. CC BY-SA license.

Figure 3. DNA structure. "DNA Replication Split." Image by Madeleine Price Ball (Madprime), 2013, Wikimedia Commons. CCO 1.0 license.

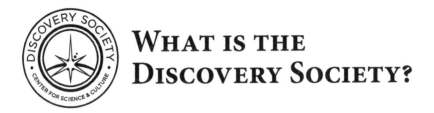

WHAT IS THE DISCOVERY SOCIETY?

The Discovery Society is a group of individuals who come together to support the work–and disseminate the message–of Discovery Institute's Center for Science and Culture. New members receive materials that help educate themselves and spread the word about our work to those in their circle of influence. Depending upon their giving level, members receive one to three Discovery Institute Press newly released books per year, along with invitations to regional donor events and discounted rates on our annual Insiders Briefing events.

If you appreciate this booklet and aren't already a member, we hope you will consider joining our network of supporters today!

Your donation to Discovery Institute's Center for Science and Culture will allow us to expand our cutting-edge scientific research and scholarship; train young people through our education and outreach; and reach the masses through media and communications.

discovery.org/id/donate

MORE INFORMATION ON THE DISCOVERY SOCIETY CAN BE FOUND AT
discovery.org/id/donate/#member-levels.

EVOLUTION AND INTELLIGENT DESIGN IN A NUTSHELL

Are life and the universe a mindless accident—the blind outworking of laws governing cosmic, chemical, and biological evolution? That's the official story many of us were taught somewhere along the way. But what does the science actually say? Drawing on recent discoveries in astronomy, cosmology, chemistry, biology, and paleontology, *Evolution and Intelligent Design in a Nutshell* shows how the latest scientific evidence suggests a very different story.

"accessible, informative… powerful … an excellent resource."

J. Warner Wallace

PURCHASE THE FULL BOOK HERE:

DiscoveryInstitutePress.com/EvolutionandID

MORE IN THIS SERIES:

This series of booklets was created to help Discovery Society members educate themselves about the basic arguments for intelligent design and the critiques of Darwinian evolution. Each booklet presents the content of one chapter of *Evolution and Intelligent Design in a Nutshell*. To help you delve deeper into each subject, we have included a list of recommended resources from our vast library of videos, podcasts, articles, and websites. Members of the Discovery Society can download digital versions of these books through the Discovery Society Community on the DiscoveryU.org platform or purchase physical copies at a discounted rate through Amazon.com.